Earth's Changing Surface
The Rock Cycle

by Kate Boehm Jerome

Table of Contents

Develop Language 2

CHAPTER 1 Breaking Down Rocks 4
            Your Turn: Communicate 9
CHAPTER 2 Grouping Rocks 10
            Your Turn: Summarize 15
CHAPTER 3 The Rock Cycle 16
            Your Turn: Interpret Data 19

Career Explorations 20
Use Language to Give Examples 21
Science Around You 22
Key Words 23
Index 24

DEVELOP LANGUAGE

A mountain of rock seems like it will last forever. But even solid rock changes over time. Rocks break down. Rocks even change from one kind of rock to another.

Look at the small pictures. Describe what you see.

The shoreline is changing because of _____.

The ice and snow of the _____ is changing the mountain.

The large rock _____ into two smaller rocks.

What else do you know about how rocks change? Share your ideas.

shoreline

glacier

El Capitan, Yosemite National Park, California

Develop Language 3

CHAPTER 1

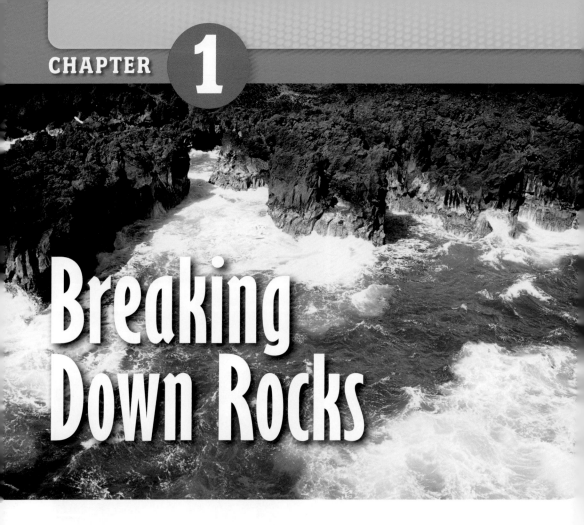

Breaking Down Rocks

Look at the rocks in the picture. Every day, ocean waves pound against them. Over time, these rocks will get smaller because of **weathering** and **erosion**.

Weathering happens when rocks break down into smaller and smaller pieces. Erosion moves these small pieces from one place to another.

weathering – how rocks break down and change
erosion – the movement of small rocks and other particles from one place to another

Weathering and erosion are caused by forces. Gravity, wind, water, and ice break down rocks or move small pieces of rock around. This changes Earth's surface.

▲ Gravity pulls things together. Earth's gravity can pull rocks down a hill.

▲ Wind carries small pieces of rock, such as sand, that can break down rocks.

▲ Water slowly wears down rocks. It can then carry smaller pieces of rock away.

▲ Glaciers are huge masses of ice. As they move, they wear down rocks and move them.

KEY IDEA Earth's surface changes because of forces such as gravity, wind, water, and ice.

Chapter 1: Breaking Down Rocks

Different Kinds of Weathering

Weathering can change rocks in different ways. **Mechanical weathering** changes the physical size of rocks. It does not change the composition, or chemical make up, of rocks.

Have you ever seen a tree growing in a rock? The tree's roots grow into small cracks in the rock. As the roots grow larger, they push on the rock. This can make the rock split. This is mechanical weathering.

mechanical weathering – the breaking down of a rock into smaller pieces without changing its composition

Tree roots can cause mechanical weathering.

Chemical weathering breaks down rocks by changing the composition of the rocks. For example, sometimes **rust** forms in a rock. This happens when oxygen from the air combines with a mineral called iron.

Rust weakens the rock. The rock can then split into smaller pieces. This is an example of chemical weathering. The rock is broken down because of changes in the composition of the rock.

chemical weathering – the breaking down of a rock because of changes in its composition

rust – a material that forms when oxygen and iron combine

KEY IDEA Weathering breaks down rocks in different ways.

By The Way...
When harmful gases in the air combine with water, acid rain forms. Acid rain causes chemical weathering. This statue is being worn away by acid rain.

◀ **Rust gives these rocks their red color.**

Chapter 1: Breaking Down Rocks   7

Weathering Produces Soil

Over a long time, the tiny pieces of rock that result from weathering begin to form **soil**. Soil is a material in the top layer of Earth's crust.

Soil contains tiny pieces of weathered rock, air, water, and **humus**. Humus forms from the remains of dead plants and animals. Living things can be found in soil, too.

The drawings on this page show how soil forms. It's a process that takes a very long time.

soil – a material in the top layer of Earth's crust

humus – a part of soil that forms from dead plants and animals

KEY IDEA Soil contains weathered rocks, air, water, and humus.

Soil Formation

▲ Rocks begin to break up.

▲ Weathered rocks mix with humus, air, and water.

▲ Living things help make the soil better.

YOUR TURN

COMMUNICATE

Make a chart like the one below. Fill in the blanks.
Write or draw an example of each type of weathering.

Weathering Process	Example
Mechanical weathering changes rocks by ____.	
Chemical weathering changes rocks by ____.	

MAKE CONNECTIONS

Spring rains can loosen weathered rocks and soil on a mountain. Why do you think drivers on mountain roads have to be extra careful at this time?

USE THE LANGUAGE OF SCIENCE

What happens in chemical weathering?

Chemical weathering breaks down rocks by changing the composition of the rocks.

Chapter 1: Breaking Down Rocks

CHAPTER 2

Grouping Rocks

Think of all the rocks on Earth. There are too many to count. But did you know scientists divide all rocks into three main groups?

The three groups are **igneous rock**, **sedimentary rock**, and **metamorphic rock**. A rock is classified into one of these groups by the way it was formed.

igneous rock – rock formed when hot, melted rock cools

sedimentary rock – rock formed when tiny pieces of rock and other particles get squeezed together

metamorphic rock – rock formed when extreme heat and pressure change one type of rock into another

Igneous rock comes from hot, melted rock below Earth's surface. This melted rock is called **magma**. Sometimes magma rises toward Earth's surface and begins to cool. It then hardens into igneous rock.

Igneous rock can form underground. It can also form above ground. When melted rock reaches Earth's surface through a volcano, it is called **lava**. As lava cools and hardens, it forms igneous rock.

magma – hot, melted rock under Earth's surface

lava – melted rock that reaches Earth's surface

Explore Language

Related Words

igneus (Latin) = of fire, from fire

igneous rock = rock formed when hot, liquid rock cools

ignite = to start burning

lava

Sedimentary Rocks

Sedimentary rock forms from weathered rocks. Forces, such as wind and water, carry away tiny pieces of weathered rock and other particles and drop them in new places. The pieces, called **sediments**, pile up in layers.

As the layers build up, the top layers of sediments squeeze the bottom layers. After a long time, this can cause the bottom layers to stick together. Sedimentary rock can form.

sediments – tiny pieces of rock and other particles that are carried to new places by forces such as wind and water

SHARE IDEAS Compare how igneous and some sedimentary rocks form.

▼ This cliff in Utah has layers of sedimentary rock.

Fossils Tell A Story

Sedimentary rock sometimes contain clues to life in the past. **Fossils** are the remains or signs of living things from long ago.

When living things die, their remains usually decay, or break down. But sometimes sediments quickly cover the remains. Then the remains may be **preserved** for a long time. That is why layers of sedimentary rock are often good places to find fossils.

fossils – the remains or signs of life in the past
preserved – saved or protected

This fossil of a Tyrannosaurus rex dinosaur was found in South Dakota.

Chapter 2: Grouping Rocks 13

Metamorphic Rocks

Deep underground, extreme heat or pressure can change the structure and composition of rocks. This changes the rocks into a different type of rock.

When one type of rock changes into another type of rock, metamorphic rock is formed. Metamorphic rocks can form from igneous, sedimentary, or even other metamorphic rocks.

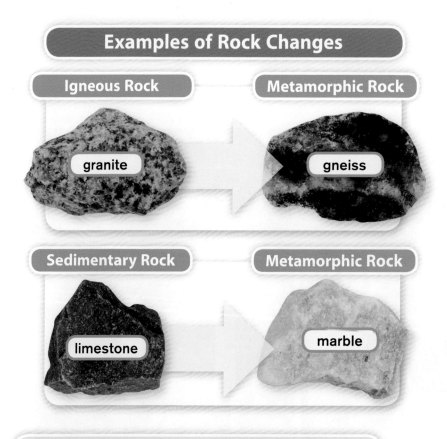

KEY IDEAS Rocks are grouped by the way they form. The three groups of rock are igneous, sedimentary, and metamorphic.

YOUR TURN

SUMMARIZE

Look back through this chapter on grouping rocks. Complete the sentence about each type of rock.

When melted rock cools, it hardens into _____ rock.

Sediments pile up. They squeeze together to form _____ rock.

Extreme heat or pressure can change rock. _____ rock forms this way.

MAKE CONNECTIONS

Scientists have found many fossils in the Badlands of South Dakota. What type of rock do you think is found there?

Strategy Focus: Synthesize

Reread the ideas on page 14 and look at the rocks. Add what you already know about rocks. Make one statement that includes most of the information.

CHAPTER 3
The Rock Cycle

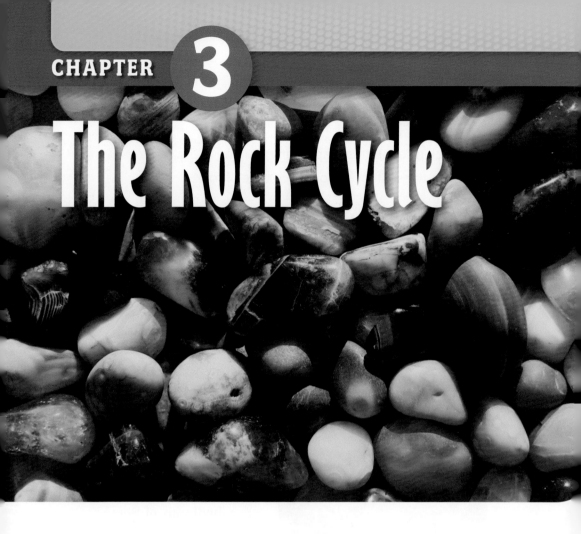

Rocks are always changing from one type to another in a never-ending cycle called the **rock cycle**. This means that rocks you see today were probably different a long time ago. The rocks will keep changing in the future. The rock cycle is a long, slow process that never stops.

> **rock cycle** – the process that happens over a long period of time in which one type of rock changes into another type of rock

In the rock cycle any type of rock can change into another type of rock. For example, under great heat and pressure, igneous rock can change into metamorphic rock. But if metamorphic rock is pushed deep within Earth it can melt. The magma can then cool to form igneous rock. Look at the diagram below to see how rocks can change into different types.

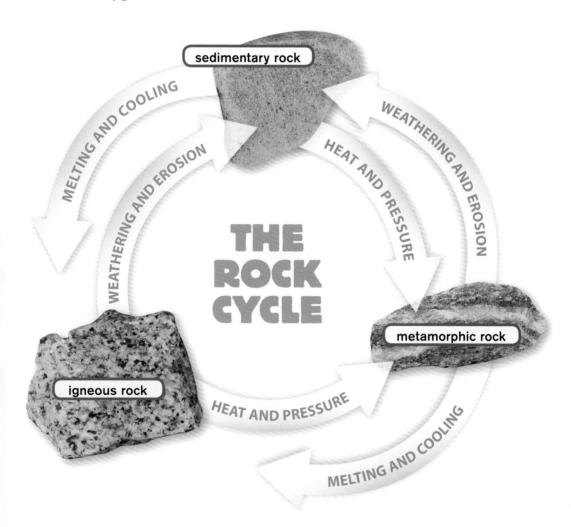

Chapter 3: The Rock Cycle 17

Recycling

Do you **recycle** material at your school or home? When you recycle aluminum cans, they can be turned into other aluminum products.

Think of the rock cycle as one of Earth's recycling programs. Older rocks are constantly changed to make younger rocks. These rocks then start to change and the cycle starts all over again.

recycle – to reuse

> **KEY IDEA** Over a long period of time, rocks can change from one type to another in the rock cycle.

YOUR TURN

INTERPRET DATA

Rocks were collected in three different areas. Look at the chart and answer the questions.

Rock	Number Found		
	Area A	Area B	Area C
Metamorphic	4	8	2
Sedimentary	12	2	0
Igneous	2	9	3

Where were most sedimentary rocks found? Area _____

Where were the fewest sedimentary rocks found? Area _____

Which area had the most metamorphic and igneous rocks? Area _____

MAKE CONNECTIONS

Water on Earth moves from oceans, lakes, and rivers into the air. Then it falls in such forms as rain or snow. How is this process like the rock cycle?

EXPAND VOCABULARY

Mechanical is related to machines, tools, or physical forces. See how **mechanical** is used in these sentences:

- A **mechanical** pencil never needs sharpening.
- **Mechanical** weathering broke the rock in two.
- Since he has **mechanical** skills, he fixed the watch.
- The candy is made using a **mechanical** process.

Compare these meanings. Tell if **mechanical** relates to machines, tools, or physical forces.

Chapter 3: The Rock Cycle

CAREER EXPLORATIONS

Do you like learning about rocks?
Find out about these careers.

If you like...	then find out about:
studying the kinds of rocks in an area	geologists
drilling and digging through rocks	well drillers
studying the composition of rocks	geochemists

▼ Geologists study rocks and changes in the surface of Earth.

▼ Well drillers dig deep to search for things like water and oil.

◀ Geochemists identify and study oil, coal, and gas fields.

20 *Earth's Changing Surface: The Rock Cycle*

USE LANGUAGE TO GIVE EXAMPLES

Words that Give Examples

When you state an important idea, you can give examples to support it. Use words like **such as** or **include**.

EXAMPLES

Earth's surface changes because of forces **such as** gravity, wind, water, and ice.

Different types of rock **include** igneous, sedimentary, and metamorphic.

With a friend, talk about soil. Give examples of different things found in soil.

Write Using Examples

Weathering breaks down or changes rocks. You learned about mechanical weathering and chemical weathering. Choose one to write about. Explain what it is, and then give examples of that type of weathering.

Words You Can Use	
such as	which is
for example	include

SCIENCE AROUND YOU

A New View Of An Old Place

Over millions of years, the Colorado River cut a deep canyon through the rocks in Arizona. Much of the Grand Canyon is now a National Park.

The Hualapai Indians own land west of Grand Canyon National Park. They built a deck for people to view the canyon. The bottom of the deck is made of glass so people can see 1220 meters (about 4000 feet) down!

What do you think you would see through the glass floor?

I would see _____ and _____ .

What river would you see at the bottom?

I would see the _____ River.

Key Words

chemical weathering the breaking down of a rock because of changes in its composition
Acid rain can cause **chemical weathering**.

erosion the movement of small rocks and other particles from one place to another
Beach **erosion** can be caused by ocean waves.

fossil (fossils) the remains or signs of life in the past
A **fossil** can give clues to life long ago.

humus a part of soil that forms from dead plants and animals
Plants grow well in soil that is rich in **humus**.

igneous rock (igneous rocks) rock formed when hot, melted rock cools
Igneous rock can form above or below the surface of Earth.

magma hot, melted rock under Earth's surface
Underground igneous rock forms when hot **magma** cools and hardens.

mechanical weathering the breaking down of a rock into smaller pieces without changing its composition
Fast-moving water can cause **mechanical weathering**.

metamorphic rock (metamorphic rocks) rock formed when extreme heat and pressure change one type of rock into another
Any type of rock can become **metamorphic rock**.

rock cycle the process that happens over a long period of time in which one type of rock changes into another type of rock
Rocks are always changing because of the **rock cycle**.

sedimentary rock (sedimentary rocks) rock formed when tiny pieces of rock and other particles get squeezed together
Sedimentary rock sometimes contains fossils.

soil a material in the top layer of Earth's crust
Weathered rock, air, water, and humus are part of **soil**.

weathering how rocks break down and change
Wind and water cause **weathering**.

Index

chemical weathering 7, 9	**magma** 11
erosion 4–5	**mechanical weathering** 6, 9
fossils 13, 15	**metamorphic rock** 10, 14–15, 17, 19
glaciers 5	**rock cycle** 16–19
gravity 5	**rust** 7
humus 8	**sediments** 12–13
igneous rock 10–11, 14–15, 17, 19	**sedimentary rock** 10, 12–15, 17, 19
iron 7	**soil** 8–9
lava 11	**weathering** 4–8

EDITORIAL LEADERSHIP TEAM AND KEY CONTRIBUTORS
Ericka Markman, Karen Peratt, Lisa Bingen, David Willette, Rachel L. Moir, Shelby Alinsky, Mary Ann Mortellaro, Amy Sarver, Betsy Carpenter, Guadalupe Lopez, Kris Hanneman and Pictures Unlimited

PROGRAM AUTHORS
Mary Hawley; Program Author, Instructional Design
Kate Boehm Jerome; Program Author, Science

BOOK DESIGN
Steve Curtis Design

CONTENT REVIEWER
Tom Nolan, Operations Engineer, NASA Jet Propulsion Laboratory, Pasadena, CA

PROGRAM ADVISORS
Scott K. Baker, PhD, Pacific Institutes for Research, Eugene, OR
Carla C. Johnson, EdD, University of Toledo, Toledo, OH
Margit McGuire, PhD, Seattle University, Seattle, WA
Donna Ogle, EdD, National-Louis University, Chicago, IL
Betty Ansin Smallwood, PhD, Center for Applied Linguistics, Washington, DC
Gail Thompson, PhD, Claremont Graduate University, Claremont, CA
Emma Violand-Sánchez, EdD, Arlington Public Schools, Arlington, VA (retired)

TECHNOLOGY
Arleen Nakama, Project Manager
Audio: Heartworks International, Inc.
Website and Online Resources: Red 7
Graphics Design: Becky Rehder, Bijou Graphics and Design

PHOTO CREDITS
Cover © Arctic-Images/Iconica/Getty Images; 1 © Simon Fraser/Photo Researchers, Inc.; 2-3 © Jamie and Judy Wild/Danita Delimont; 2 © Value Stock Images/age fotostock; 3a © R. Matina/age fotostock; 3b © Simon Fraser/Photo Researchers, Inc.; 3c © SuperStock/age fotostock; 4 © Photodisc/Punchstock; 5a © AP Images/The Wayne County News/Brian Sizemore; 5b © Tom Bean/Getty Images; 5c © Kenneth Sponsler/Shutterstock; 5d © Robert Harding Picture Library/Photolibrary; 6 © Ron Adcock/Alamy; 7a © Prosoco; 7b © Marco Brivio/Alamy; 8 illustration by Sharon and Joel Harris; 9a © Harris Shiffman/Shutterstock; 9b and 9c Lloyd Wolf for Millmark Education; 10, 14a, 14b, 17a, 17b, 17c © Wally Eberhart/Visuals Unlimited; 11 © G. Brad Lewis/age fotostock; 12 © David R. Frazier/Photo Researchers, Inc.; 13 © David R. Frazier/Danita Delimont; 15 © Will Hart/PhotoEdit; 16 © Mack7777/Shutterstock; 18a © prism 68/Shutterstock; 18b © Lonnie Duka/Index Stock Imagery; 18c © Studio Araminta/Shutterstock; 20a © Greenshoots Communications/Alamy; 20b © AP Images/The Denver Post/R.J. Sangosti; 20c © AP Images/Andrew Rush; 21 © gary718/Shutterstock; 22 © David Kadlubowski/Corbis; 24 © Barbara Jablonska/Shutterstock

Copyright ©2016 Summit K12 Holdings, Inc.

All rights reserved. Reproduction of the whole or any part of the contents without written permission from the publisher is prohibited. Millmark Education and ConceptLinks are registered trademarks of Summit K12 Holdings, Inc.

Published by Summit K12 Holdings, Inc.
PO Box 92421
Southlake, TX 76092

ISBN-13: 978-1-4334-0071-1

Printed in the USA

10 9 8 7 6 5 4 3